JAN 3 1 2022

How Do You Live There?

LIVING IN SPACE

Carol Hand

PowerKiDS
press

NEW YORK

Published in 2021 by The Rosen Publishing Group, Inc.
29 East 21st Street, New York, NY 10010

Editor: Kristen Susienka
Designer: Rachel Rising

Photo Credits: Cover, Paopano/Shutterstock.com; pp.1,4,6,8,9,10,12,14,15,16,18,20,21,22,24,25,26,27,28,29, 30,31,32 (background) Yevhenii Borshosh/Shutterstock.com; p. 5 BestStockFoto/Shutterstock.com; p. 7 Albert Barr/ Shutterstock.com; p. 9 Vladi333/Shutterstock.com; p. 11 Popperfoto / Contributor/Getty Images; pp. 13,17 Bettmann / Contributor/Getty Images; p.14 SAUL LOEB/AFP via Getty Images; p. 15 NASA/ NASA History; p. 19 Science & Society Picture Library / Contributor/Getty Images; p. 20 3Dsculptor/Shutterstock.com; p. 21 Det-anan/Shutterstock.com; p. 23 Andrey Armyagov/Shutterstock.com; p. 25 NASA / Handout/Getty Images; pp. 26,29 Pavel Chagochkin/Shutterstock. com; p .27 Naeblys/Shutterstock.com; p. 28 e71lena/Shutterstock.com; p. 30 Castleski/Shutterstock.com.

Some of the images in this book illustrate individuals who are models. The depictions do not imply actual situations or events.

Library of Congress Cataloging-in-Publication Data

Names: Hand, Carol, 1945- author.
Title: Living in space / Carol Hand.
Description: New York : PowerKids Press, [2021] | Series: How do you live
    there? | Includes index.
Identifiers: LCCN 2019050474 | ISBN 9781725316638 (paperback) | ISBN
    9781725316652 (library binding) | ISBN 9781725316645 (6 pack)
Subjects: LCSH: Life support systems (Space environment)--Juvenile
    literature. | Space environment--Juvenile literature. | Space
    flight--Juvenile literature. | Space stations--Juvenile literature. |
    Outer space--Exploration--Juvenile literature.
Classification: LCC TL1500 .H36 2021 | DDC 629.45--dc23
LC record available at https://lccn.loc.gov/2019050474

Manufactured in the United States of America

CPSIA Compliance Information: Batch #CSPK20. For further information contact Rosen Publishing, New York, New York at 1-800-237-9932.

Find us on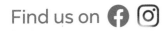

CONTENTS

WHAT IS SPACE? 4

WHAT IS SPACE MADE OF? 6

LIGHT AND SOUND 8

HUMANS AND SPACE 10

THE CHALLENGES OF SPACE 12

LIFE WITHOUT GRAVITY 14

EARLY ROCKETS AND SPACECRAFT 16

A TRIP TO THE MOON 18

STATIONS AND SHUTTLES 20

HOW THE ISS WORKS 22

LIVING ON THE ISS 24

A MOON COLONY 26

TO MARS AND BEYOND 28

BIG MOMENTS IN AMERICAN SPACE EXPLORATION 30

GLOSSARY 31

INDEX 32

WEBSITES 32

WHAT IS SPACE?

Looking up from Earth's surface, the sky looks blue. This is because air in the **atmosphere** reflects light. Higher up, there is less air. As you get even higher, all the air goes away. The sky turns from blue to black. This is outer space.

Scientists disagree about exactly where outer space begins. Many think that space starts at a place called the Kármán Line. It is 62 miles (100 kilometers) above Earth's surface. The atmosphere is too thin for airplanes to fly above this height. The National Aeronautics and Space Administration (NASA) thinks space begins lower. They say it starts 50 miles (80 km) above Earth's surface.

Some argue that space begins where the highest layer of Earth's atmosphere ends. That's about 6,200 miles (10,000 km) high!

The highest layer of Earth's atmosphere has almost no air **particles**. Most people think of it as part of outer space. In fact, the International Space Station (ISS) flies below this layer of the atmosphere.

WHAT IS SPACE MADE OF?

No one knows how big space is. It contains many large objects like stars and planets. The sun is a star. The sun and its planets form our solar system. The planets orbit, or circle, the sun. The sun's **gravity** pulls planets toward it and holds them in orbit.

A **galaxy** has billions of stars held near each other by gravity. Our sun and solar system are part of the Milky Way galaxy. There are hundreds of billions of galaxies in the **universe**.

Even with all the planets, stars, and galaxies, most of space is almost completely empty. Space is a **vacuum**. Open space has almost no matter, except some gas and dust particles. These particles are very far apart compared to the air particles in Earth's atmosphere.

Spiral galaxies such as this one and the Milky Way contain billions of stars held in place by gravity.

LIGHT AND SOUND

There is no sound in space. Sound needs a medium, like air or water, to travel through. Sound travels by making the particles in the medium bump into each other. Because space is nearly empty, there aren't enough particles for sounds to spread.

Space has radiation, or energy released in waves. Light is an example of radiation. Light doesn't need a medium to spread. It can move through empty space.

Light moves very quickly—about 186,000 miles (300,000 km) every second—but it still takes time for light to get from place to place. Objects in space are very far apart. This is why scientists often use a unit called light-years to measure distances in space. A light-year is the distance light travels in a year. One light-year equals about 5.8 trillion miles (9.3 trillion km)!

Light from the sun takes about eight minutes to reach Earth. ⊢⟶

A Look Back in Time

For centuries, people have looked to the North Star to help find their way. That star, also called Polaris, is 323 light-years away from Earth. That means the light we see when we look at it was actually created 323 years ago! The light from the most distant star that scientists have been able to see took 9 billion years to reach Earth. Looking deep into space is like looking back in time.

HUMANS AND SPACE

For as long as humans have been alive, they've been interested in space. Ancient cultures studied the night sky to learn more about the universe. By watching the sky, the ancient Greeks figured out the size of Earth. The Maya civilization of Central America made calendars that said when **eclipses** would happen hundreds of years in the future.

In the early 1600s, people began creating telescopes. As telescopes got better over time, our understanding of planets and stars improved.

Beyond studying space from the ground, humans have always wanted to live there. But getting to space is difficult. Space is also very dangerous. Keeping people alive there comes with many challenges. Today's scientists and engineers have solved some of these problems. Others still need to be solved before humans can live their whole lives in space.

Galileo Galilei lived in Italy from 1564 to 1642. The telescopes he built helped him study the moon and the planets.

GALILEO GALILEI

THE CHALLENGES OF SPACE

Earth's gravity is very strong. It pulls things toward Earth's center. To get into outer space, people must build powerful rockets. A rocket must fly at 25,000 miles (40,230 km) per hour or more to break free of Earth's gravity.

People cannot live in the vacuum of space. They must have air to breathe. They need to stay inside a spacecraft or spacesuit. Space also lacks other things people need for life, like food and water. Space travelers must bring their own special supplies from Earth.

Temperature is another challenge in space. Some parts of space are very hot. Others are very cold. The part of the moon facing the sun can be 260°Fahrenheit (127°Celsius). The part facing away from the sun can be −280°F (−173°C). People need protection from these **extremes**.

Outside Earth's atmosphere, radiation is deadly. Radiation damages the brain and causes cancer. People in space must be kept safe from radiation.

Spacecraft are launched attached to huge rockets, as shown here. Rockets burn a lot of fuel to go fast enough to reach outer space.

LIFE WITHOUT GRAVITY

Without gravity pulling on them, people and objects in space are weightless. When they are in space, people can float, zoom around, and lift heavy objects easily. They can walk on the walls, ceiling, or floor. Their tools float around them too.

This may sound like fun, and it definitely can be. But being weightless also has its costs. Without gravity, blood does not run down toward the feet. Faces swell, and noses get stuffy. Without the force of gravity pressing on their bodies, people's bones become weak. Legs become thin, and muscles get weak and flabby. Bones in the spine separate, and people have back aches. The longer someone spends in space, the more serious these problems can become. Worst of all, every day in space is a bad hair day!

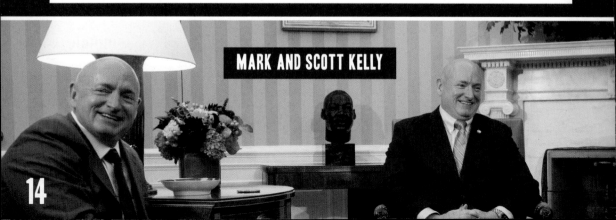

MARK AND SCOTT KELLY

NASA's Twin Astronauts

Scientists have many questions about the health effects of life at zero gravity. Luckily, two NASA astronauts are identical twins (see page 14 photograph). In 2015, Scott Kelly spent almost a year on the International Space Station, while his twin brother, Mark, remained on Earth. By comparing Scott and Mark's bodies, scientists learned much about how space changes the human body. This has helped them prepare future astronauts for long-term space living.

Without gravity, a **spacewalk** is really a "space float." To keep from drifting away, astronauts travel in space using a special machine.

EARLY ROCKETS AND SPACECRAFT

The first successful rockets were built in the mid-1900s. They were powerful enough to escape the pull of Earth's gravity. Spacecraft are attached to rockets and sent into space. When they get high enough, the two come apart, and the spacecraft continues alone.

Spacecraft designers have a tough job. Even a tiny crack could allow the air to be sucked out into space. Spacecraft need to be strong enough to survive being launched. But they can't be too heavy, or they won't get off the ground.

The first American in space was Alan Shepard. His spacecraft, the *Freedom 7*, was small and cramped. It was made mostly from titanium and steel. These metals are strong but lightweight. *Freedom 7* could protect against extreme heat as it returned to the atmosphere.

Alan Shepard became the first American in space on May 5, 1961. His flight was part of NASA's Mercury program.

17

A TRIP TO THE MOON

Humans began going to space in the early 1960s. These first trips stayed close to Earth. As rockets got stronger and spacecraft got better, missions became more advanced. On July 16, 1969, the United States sent the Apollo 11 mission to the moon. It arrived on July 20.

Astronauts Neil Armstrong and Buzz Aldrin became the first people to walk on the moon. They wore special spacesuits. They were made from 25 layers of material for protection from radiation and extreme heat and cold. They had to be flexible enough for the astronauts to move around, but strong enough not to break.

The Apollo 11 mission was powered by fuel cells. Like batteries, the fuel cells made electricity by causing a **chemical reaction**. This reaction also produced water, which the astronauts could drink.

Astronauts Neil Armstrong (*left*) and Buzz Aldrin (*right*) practice moving in their spacesuits and gathering rock samples a few months before their 1969 mission to the moon. ⊢⟶

Space Food

On the first space missions, food was kept in tubes, like toothpaste. Astronauts could squeeze it directly into their mouths. Later, food was freeze-dried and sealed in air-tight packages. This took most of the water out of the food, making it very light. It could last for a long time without being refrigerated. Today, space food is improving. Astronauts are even learning to grow fresh fruits and vegetables in space.

STATIONS AND SHUTTLES

The Apollo 11 mission to the moon lasted nine days. More recently, scientists have wanted to carry out longer missions in space. To do this, they've built several space stations, put them in orbit around Earth, and sent people to live in them. Early space stations included NASA's Skylab and Russia's Mir. Today, astronauts live on the International Space Station (ISS).

Space shuttles could fly in space and land back on Earth.

This is the view from inside a space shuttle cockpit on display at Space Center Houston in Houston, Texas.

From 1981 through 2011, the United States ran the space shuttle program. Like earlier spacecraft, space shuttles were launched by powerful rockets. Unlike earlier craft, however, they could land on a runway and be used again. Shuttles had crews of five to seven astronauts. Flights lasted up to 17 days. Shuttles could dock with space stations to deliver people and supplies. Together, NASA's fleet of five space shuttles flew 135 missions.

HOW THE ISS WORKS

The International Space Station was launched section by section and put together in space. The first section was launched in 1998. The ISS was finished in 2011, though changes are still being made.

To make oxygen, astronauts use electricity to split water into oxygen gas and hydrogen gas. The electricity comes from solar panels. Carbon dioxide that people breathe out is pulled from the air and sent into space.

Temperature control is also necessary on the ISS. It is covered in material that reflects light. This keeps the station from getting too hot in the sun. Inside the ISS, however, people and machines also produce a lot of heat. A cooling system uses pipes filled with cold liquid to soak up extra heat and release it into space.

LIVING ON THE ISS

Astronauts have lived on the ISS constantly since 2000. Most come from the United States or Russia. There are three to six astronauts on the ISS at a time. Each usually stays for about six months.

Astronauts do science experiments. They keep the equipment clean and working properly. Sometimes, they take spacewalks to repair things outside the ISS. They exercise using a treadmill and a bicycle. Astronauts use these machines by wearing special equipment. There's also a machine that creates the experience of weightlifting.

Astronauts spend their free time reading, listening to music, or doing other hobbies. They love looking out the space station windows at the beautiful views of Earth. They sleep in sleeping bags attached to the floor, ceiling, or wall. This keeps them from floating around.

Spacewalks can be tiring. Here, astronaut Scott Kelly repairs a cooling system on the ISS. ⊢——————→

Space Toilets

Toilets on the ISS are small. The user must aim very carefully. Waste is carried away by suction instead of water and gravity. The poop is stored and sent back to Earth on the next cargo ship. It burns up in the atmosphere, like shooting stars. The pee is cleaned and turned into water. Some is used for drinking, and some is used to make oxygen inside the ISS.

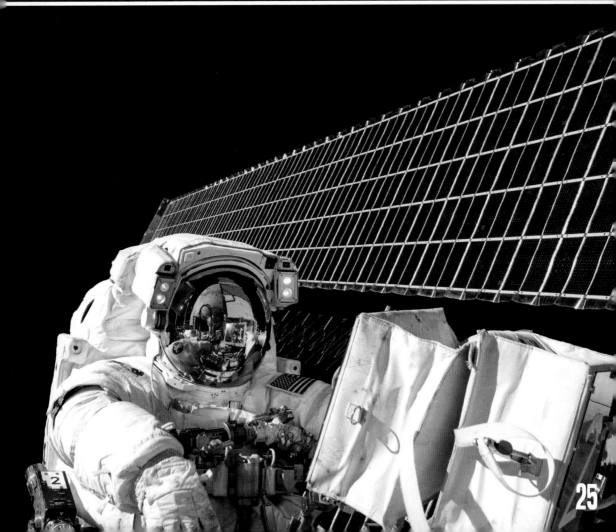

A MOON COLONY

Outer space is a new and exciting place to explore. Scientists want to better understand space itself. They want to learn more about planets and moons.

People also want **resources** from space. Asteroids have metals that can be mined and brought back to Earth. A 2015 study suggested that a mining town could be built on the moon. People would mine ice from the poles and use it to make hydrogen gas for fuel. Robots would help find resources to mine.

People imagine what life on Mars and other planets would be like.

Just as we look up and see the moon at night, colonists on a moon base would look up and see Earth.

Many people want to **colonize** the moon. Through NASA's Artemis program, people will return to the moon in 2024. The goal is to set up a permanent human presence there, similar to the ISS. Private companies like SpaceX will help with this project.

Having a colony on the moon would teach people about living in space. It would help prepare them for colonizing places even farther away. NASA hopes its Artemis program will get humans ready to go to Mars in the 2030s.

Many scientists want to form camps on Mars where people could live and do research for a short time. These outposts could later become colonies where people would live permanently.

A company called Mars One wants to start a colony on Mars as soon as possible. In 2013, they started looking for volunteers for one-way missions to Mars. That means the people would go to Mars for the rest of their lives. Would you want to volunteer for a one-way mission to Mars?

Space Junk

Space is full of junk. There are more than 20,000 objects orbiting Earth that are larger than a softball, including dead **satellites** and old rocket parts. Even a fleck of paint can cause damage if it slams into a passing spacecraft. Scientists want to think more carefully about what gets left behind during space missions—not just to stop crashes, but also to keep space clean for future visitors to enjoy.

Some scientists imagine people living in domes on Mars, with forests growing inside them. The forests would make oxygen for the people to breathe.

BIG MOMENTS IN AMERICAN SPACE EXPLORATION

May 5, 1961 Alan Shepard becomes the first American in space.

February 20, 1962 John Glenn becomes the first American to orbit Earth.

July 20, 1969 Neil Armstrong and Buzz Aldrin become the first humans to set foot on the moon.

December 14, 1972 Harrison Schmitt and Eugene Cernan leave the moon after spending a record 75 hours there. No human has returned to the moon since.

May 14, 1973 NASA launches Skylab, the first American space station.

June 18, 1983 Aboard the space shuttle *Challenger*, Sally Ride becomes the first American woman in space.

August 30, 1983 Guion Bluford becomes the first African American in space.

November 2, 2000 The ISS welcomes its first crew, including American astronaut William Shepherd.

March 27, 2015 Scott Kelly begins his yearlong stay on the ISS as part of NASA's twin study.

September 3, 2017 Peggy Whitson returns to Earth after spending 289 days in space, a record for an American woman.

October 18, 2019 Christina Koch and Jessica Meir conduct the first all-female spacewalk.

GLOSSARY

atmosphere: All the gases surrounding Earth.

chemical reaction: Something that happens when different substances are combined.

colonize: To send a group of people to a new place to live permanently.

eclipse: When one body in space passes in front of another, such as the moon passing in front of the sun.

extreme: Too much of something, such as heat or cold.

galaxy: A group of billions of stars held together by gravity.

gravity: The force that attracts any body in space to another.

particle: The smallest possible amount of something.

resource: A material like metal or gold that can be used on Earth to make things.

satellite: An object that people send into orbit.

spacewalk: Floating, or "walking," around the outside of a space station or spacecraft, usually to make repairs.

universe: All the stars, planets, and galaxies, and the space around them.

vacuum: An area that is completely empty, with no matter.

INDEX

A
air, 4, 5, 6, 8, 12, 16, 22, 25, 29
Aldrin, Buzz, 18, 30
ancient cultures, 10
Apollo 11 mission, 18, 20
Armstrong, Neil, 18, 30
Artemis program, 27, 28
atmosphere, 4, 5, 6, 12, 16, 25

C
colonies, 26–27, 28
cooling system, 22, 24

E
exercise, 24
experiments, 15, 24

F
food, 12, 19

G
galaxies, 6, 7
Galilei, Galileo, 10
gravity, 6, 7, 12, 14, 15, 16, 25

H
health, 14, 15

I
International Space Station, 5, 15, 20, 22, 24, 25, 27

K
Kelly twins, 14, 15, 24, 30

L
light, 4, 8, 9, 22
light-years, 8, 9

M
Mars, 26, 28, 29
moon, 10, 12, 18, 20, 26, 27, 28

N
NASA, 4, 15, 17, 20, 21, 27, 28

O
orbit, 6, 20, 22, 29, 30
oxygen, 22, 25, 29

P
particles, 5, 6, 8

R
radiation, 8, 12, 18
rockets, 12, 13, 16, 18, 21, 29

S
satellites, 29
Shepard, Alan, 16, 17, 30
sleeping, 24
solar panels, 22
solar system, 6
sound, 8
space junk, 29
space shuttles, 20, 21, 30
space stations, 20, 21

(spacesuits)
spacesuits, 12, 18
spacewalks, 15, 24, 30

T
temperature, 12, 16, 18, 22
toilets, 25

V
vacuum, 6, 12

W
water, 8, 12, 18, 19, 22, 25

WEBSITES

Due to the changing nature of Internet links, PowerKids Press has developed an online list of websites related to the subject of this book. This site is updated regularly. Please use this link to access the list:
www.powerkidslinks.com/hdylt/space